Disaster management at Health Care Settings

Comprehensive assessment and effective mitigation

Shreen Gaber

Disaster management at Health Care Settings

Content

Disaster management at Health Care Settings

First Printing: 2015

ISBN: 978-1-329-69624-2
    90000
www.lulu.com

Cairo University (FON)
Cairo University Rd, Oula, Giza, 11562
Cairo, Egypt
www.cu.edu.eg.com

Special discounts are available on quantity purchases by corporations, associations, educators, and others. For details, contact the publisher at the above listed address.

U.S. trade bookstores and wholesalers: Please contact Shreen Gaber Tel: (+20) 100-7653998; Fax: (+20) 23657190 or email shreen17@cu.edu.eg

Preface

This book draws an attention to the different type of disasters; human made and natural, in addition to provide the suggested guidelines, strategies that help in managing these disasters safely. the researcher strive to write the book in a comprehensive, concise and feasible sequential steps to be easily understood aspiring at minimize the adverse effect of disaster on health care settings and overall our community. Whether you are a beginner or an experienced health care provider, I hope that you find this book enjoyable and clinically.

Disaster management at Health Care Settings

Disaster management at Health Care

Settings *(Comprehensive assessment and effective mitigation)*

Introduction

In fact a disaster is indiscriminate in whom it affects, the health care centers usually one of the most affected public facilities as a result of dependency of stakeholders (patients) and increasing the flow of health care seekers during the event of disaster. Ardalan, Mowafi, &Yousefi, 2013 evidenced that 119 natural hazard events were recorded in last 10 years in Iran that led to physical damage and/or functional failure in 1,401 health centers, 127 deaths and injury or illness in 644 health staff. For these reasons and others the health care settings should be establish a highly measures of mitigation and preparedness to meet the different types of disaster affectively as possible. Many hospital accreditation systems such as joint commission had submitted the disasters management plan as a one of requirements to be accredited. This book aims to draw attention to the different types of disasters; human made and natural or (non preventable), in addition to provide the suggested guidelines, strategies that helps in managing these disasters. the researcher strive to write the book in a comprehensive, concise and feasible sequential steps to be easily understood aspiring at minimize the adverse effect of disaster on health care settings and overall our community.

Around the world, health care setting especially emergency departments are experiencing unprecedented disasters from natural causes like floods, hurricanes and earthquakes to man-made disasters such as oil spills and terrorism, fire, crush motor car accidents and bioterrorism. Disasters, whether human man-made or natural, are unavoidable, but there are ways to help health care setting to prepare, respond to, and recover from disaster. So we will describe disaster management approaches including phases of prevention, preparedness, response, and recovery. The emergency front line health nurse play a significant role in these phases respectively. Baack, S. (2011)

A Disaster refers to any sudden natural or man-made (or technological) hazard resulting in an event of substantial extent causing significant damage, economic disruption, loss of human life and deterioration in health services. Disasters worldwide cost 3 million lives and affected at least 800 million others. Korstanje, 2011. Prevention is better than cure, while natural disasters cannot be prevented, but measures can be taken to eliminate or reduce the possibility of trouble and enhance the capability to survive. As well as prevention is the best protection against disaster especially man-made. The plan of management should be at the level of institutional, governmental and community based. . John labby, (2010).

Disaster Concepts.

A disaster is any innate or man-made occurrence that causes disruption, destruction, and/or devastation requiring external support. Although natural incidents like earthquakes or hurricanes trigger many further disasters, predictable and preventable human-made factors can further affect the disaster. On August 30, 2005, the day after Hurricane Katrina hit New Orleans, a breach in the Lake Pontchartrain levees created a disaster within a disaster as 75% of the city filled with up to 20 feet of water (Reagan, 2005). From a health care standpoint, the disaster event type and timing predict subsequent injuries and illnesses. If there is prior warning (e.g., in hurricanes or earthquake), the impact brings fewer injuries and deaths. Disasters with little or no advance notice such as terrorism events will often have more casualties because those affected have little time to make evacuation preparations. Disasters with warnings also carry their own dangers, because individuals can be injured attempting to prepare for the disaster or while evacuating the affected area. So finally we can say that disasters create imperative needs across a widespread region. Crisis management workbook, (2010).

Disaster management is a process or strategy that is implemented when any type of catastrophic event takes place. Sometimes referred to as disaster recovery management, the process may be initiated when anything threatens to disrupt normal operations or puts the lives of human beings at risk.

Governments on all levels as well as many businesses create some sort of disaster plan that make it possible to overcome the catastrophic events and return to normal function as quickly as possible that closely related to and depend up on the organization facilities, resources and its manpower orientation and training. (Disaster Action Team Handbook, 2008).

One of the essential elements of disaster management involves defining the types of catastrophes that could possibly disrupt the day to day operation of the hospital or even overall country. Identifying those potential disasters makes it possible to create eventuality plans, assemble supplies, and create procedures that can be initiated when and if a given disaster does come to pass. A truly comprehensive disaster management plan will take in a wide range of possibilities that can easily be adapted in the event one disaster sets off a chain reaction of other types of disasters in its wake. Fairfax County Public Schools (2010).

Because of the need to continue functioning in emergency situations, disaster management plans are often multi-layered and can address such issues as floods, hurricanes, fires, bombings, and even mass failures of utilities or the rapid spread of disease. The disaster plan is likely to address such as important matters as evacuating people from an impacted region, arranging temporary housing, food, and medical care. It is not unusual for the plan to also work toward containing and possibly neutralizing the root Drenkard,2002)

Egypt - Disaster Statistics

Data related to human and economic losses from disasters that have occurred between 1980 and 2013.
Natural Disasters from 1980 - 2013

Overview

No of events:	23
No of people killed:	1,927
Average killed per year:	99
No of people affected:	262,864
Average affected per year	8,479
Economic Damage (US$ X 1,000):	1,342,000
Economic Damage per year (US$ X 1,000):	43,290

Effective disaster communication

The process of disaster management will often address the issue of ongoing communication. Since many disasters can cause some sort of problems of the hospital communication networks, a competent disaster management plan will include the quick setup of alternative communication capabilities that do not rely on the various switches, towers and hubs that are usually part of telephone and cellular communication networks. Often making use of short-wave transmissions that are supported with satellite technology, the communication flow can continue from the area impacted by the disaster to other points where assist from different sectors can be provided can be extended when and as possible.Eric,K. (1997),

Creating an effective disaster management plan is often easier said than done. As many nations have learned, what were thought to be comprehensive emergency plans turned out to be partially effective at best and what exactly will be done or carried out in the response phase if each type of disasters according. In recent years, many government agencies stretching from the local to the national level have taken steps to revisit the structure of their disaster plans and run computer simulations to identify weaknesses in the plans, and refine them so they can operate with more speed and efficiency, Doyle, (2009).

Attribution disaster Communication Best Practices

1. All victims or potential victims should receive instructing information, including recall information. This is one-half of the base response to a disaster.

2. All victims should be provided an expression of sympathy, any information about corrective actions and trauma counseling when needed. This can be called the "care response." This is the second-half of the base response to a crisis.

3. For crises with minimal attributions of crisis responsibility and no intensifying factors, instructing information and care response is sufficient.

4. For crises with minimal attributions of crisis responsibility and an intensifying factor, add excuse and/or justification strategies to the instructing information and care response.

5. For crises with low attributions of crisis responsibility and no intensifying factors, add excuse and/or justification strategies to the instructing information and care response.

6. For crises with low attributions of crisis responsibility and an intensifying factor, add compensation and/or apology strategies to the instructing information and care response.

7. For crises with strong attributions of crisis responsibility, add compensation and/or apology strategies to the instructing information and care response.

8. The compensation strategy is used anytime victims suffer serious harm.

9. The reminder and ingratiation strategies can be used to supplement any response.

10. Denial and attack the accuser strategies are best used only for rumor and challenge crises.

- **Notification and alerting in effective communication. Nursing Role in Disaster Communication:**

Nurses working as members of an assessment team need to return accurate information to relief managers to facilitate rapid rescue and recovery. A part of that communication is involved with the rapid and ongoing needs assessment just described. Lack of or inaccurate information regarding the scope of the disaster and its initial effects can contribute to a mismatched resource supply. After Hurricane Andrew in 1992, a well-meaning public continued to ship thousands of pounds of clothing to South Florida. Much of the clothing eventually was burned because there were inadequate on-site personnel to sort and distribute the clothing, and the piles eventually became a public health nuisance. Times of crisis or great uncertainty call for great skills in communication. The community needs accurate information transmitted in a timely manner, Dybdahl (2001).

Health care personnel are the best sources for essential health information that is technical in nature. Disaster incidents also use public affairs spokespersons for formal communication. The Public Information Officer (PIO) is an individual with the authority and responsibility to communicate information to the public at large. Still, nurses are considered trustworthy sources of information and may be approached for an interview. The nurse should refer the media to the PIO representing the agency. If the public approaches the nurse for information, however, that health information should be conveyed. It is entirely within public health nursing scope of practice to provide health education. Finally, although there are official spokespersons in all major disasters, there may be an occasion for the nurse to serve as a member of the risk communications' team. (Korstanje, M. 2011).

Risk communication

Is the "science of communicating critical information to the public in situations of high concern? The objectives in emergency communication are to identify and respond to the barriers of fear, panic, distrust, and anger: build or re-establish trust; resolve conflicts; and coordinate between stakeholders so that the necessary messages can be received, understood, acted and accepted. The first step of any response to a disaster is learning that the disaster has occurred. For many disasters, this is not a problem. The Red Cross may learn of disasters from radio and television news or from people that live in the affected area and report the incident directly. But single-family disasters rarely produce public notice, and the Red Cross must establish and maintain a network that ensures that it receives timely notification that such disasters have occurred, Attorney (2005).

Top7 to try achieving an effective management

- Triage Nurse (Job Description, definition of triage nurse, responsibilities)
- Emergency Room head nurse Job Description
- Skills Assessment
- Emergency Job Descriptions
- Hospital Patient Flow
    Improve Patient Flow Now Using Advanced Analytics
- Human Capital Management
    Improve outcomes using integrated workforce management solutions.
- Unicellinc.

Disaster management and the Role of Intuition

Researchers have paid a lot of attention to the role of intuition in decision making. They maintain that intuition can be developed and offer a variety of suggestions. Gary Klein (2003) advises learning to detect problems through emotional cues—a "gut feeling" when something isn't right. He recommends developing an active stance, so that if something doesn't make sense, it acts as an alarm that is not to be dismissed. He also suggests becoming conscious of organizational barriers such as rigid procedures or institutionalized inertia. Finally, he suggests reframing the situation and consulting with colleagues to review with fresh eyes, Dybdahl.(2001).

We can say that good decision making relies on a balance of conscious and instinctive thinking. Reducing complex problems to their simplest elements aiding in decision making. He warns that too much information can paralyze the unconscious. In other words, remember to keep it simple. improving critical-thinking skills by avoiding behaviors such as stereotyping others, resisting change, and seeking conformity. In addition, she suggests reducing barriers to intuition: anxiety, stress, fatigue, lack of time, feeling judged, and environmental distractions, Duong, (2009).

Management standards:

There are no nationally-adopted emergency management standards for the organizations, but the U.S. Department of

Education's Practical Information on Crisis Planning: A Guide for Schools and Communities, which provides a four-phase approach to school emergency planning activities, is widely used throughout the country. In addition, many different organizations have their own emergency management mandates, policies, and procedures, often based on this guide.

A number of emergency management standards intended for voluntary adoption by emergency response organizations are produced by standards-making organizations such as the National Fire Protection Association (NFPA), the American National Standards Institute (ANSI), the Institute of Electrical and Electronics Engineers (IEEE), the Organization for the Advancement of Structured Information Standards (OASIS), and ASTM International (formerly the American Society for Testing and Materials).

- National Incident Management System (NIMS):

A nationwide, comprehensive approach for addressing all hazards recommends the following voluntary standards for emergency management use.

Schools are encouraged to review these standards carefully and to adopt, where applicable, those that meet their needs. Questions about their application should be addressed to your local or state emergency response organizations.

- ASTM is currently drafting ASTM WK 8908,

Standard Guide for School Preparedness and All Hazard Response. No publication date has been announced. Refer to the "NIMS Recommended Standards List" on the NIMS Resource Center webpage. The Federal Emergency Management Agency is a part of the U.S. Department of Homeland Security.

- NFPA 1561. Standard on Emergency Service Incident Management System Overview. Defines and describes the essential elements of an incident management system that promotes coordination among responding agencies.

- NFPA 1600, Standard on Disaster/Emergency Management and Business Continuity Programs, provides a foundation for disaster/emergency management planning and operations and describes common elements, techniques, and processes.

- NFPA 1221, Standard for Installation, Maintenance, and Use of Emergency Services Communications Systems, Covers the installation, performance, operation, and maintenance of public emergency services communications systems and facilities.

- ANSI INCITS 398, Information Technology —

Common Biometric Exchange Formats Framework (CBEFF) Describes the data elements necessary for biometric data interchange among proprietary application programs.

- IEEE 1512, Standard for Common Incident Management Message Sets for Use by Emergency Management Centers Provides definitions, specific messages, data frames, and data elements for communicating information relating to traffic incidents, public safety, and hazardous cargo.

- OASIS Emergency Data Exchange Language (EDXL) Distribution Element Provides a standard message distribution structure for data sharing among emergency information systems using the XML-based emergency data exchange language.

- OASIS Common Alerting Protocol Provides a general format for exchanging all-hazard emergency alerts and public warnings, allowing a consistent warning message to be disseminated simultaneously over many different warning systems. (National Institute of Building Sciences 2009).

Types of disaster according to the hospital responsibility

Victim Crises: Minimal Crisis Responsibility

Natural disasters: acts of nature such as tornadoes or earthquakes.

Rumors: false and damaging information being circulated about you organization.

Workplace violence: attack by former or current employee on current employees on-site.

Product Tampering/Malevolence: external agent causes damage to the organization.

Accident Crises: Low Crisis Responsibility

Challenges: stakeholder claim that the organization is operating in an inappropriate manner.

Technical error accidents: equipment or technology failure that cause an industrial accident.

Technical error product harm: equipment or technology failure that cause a product to be defective or potentially harmful.

Preventable Crises: Strong Crisis Responsibility

Human-error accidents: industrial accident caused by human error.

Human-error product harm: product is defective or potentially harmful because of human error. Organizational misdeed: management actions that put stakeholders at risk and/or violate the law. (Joseph & Sandesh. 2010).

Nursing preparing for disasters

Nurses, just like other health professionals, have a critical role to play to make sure they are prepared to play an active role in disasters, both in the early stages, and in the short-and long-term recovery of communities. Since the inception of modern nurse training by Florence Nightingale in 1860, nurses across the world have been involved in disasters, particularly in times of war. As a young nurse was taught that 'Nursing was born in the church and bred in the army'. Military nurses are often an unexploited resource of knowledge and skills for those of us working in civilian settings. They have to be in a constant state of readiness for wartime and other disasters, so it makes good sense that nurses in many settings have good dialogues with them when learning about disaster preparedness. Duong. (2009).

In a number of countries there are increasing efforts to improve the education and training of nurses regarding disaster preparation. However for the most part we argue that our training and education as nurses often has not prepared us for working in emergencies beyond the walls of hospitals where there are many resources at our disposal. What happens if our hospital is destroyed or badly damaged? Or suddenly we are faced with a calamity in our district that is outside of the scope of usual practice? Clearly we need to cross-examine our nursing curriculum and our practices to ensure that all nurses have better knowledge and skills in disaster preparation and recruitment. The recently developed World Health Organization (WHO) International Council of Nurses (ICN) Disaster Nursing Competencies should help because they provide a clear framework for the work of nurses in disasters and would be useful in programs and short courses in nursing staff. (Uscher Pines. 2009).

The latest sort of disasters in our Egyptian hospitals highlights the critical need for enlistment of aid, for early warning systems, and advanced planning across a wide range of instrumentalities, government and non-government, especially in hospitals that are poor where available infrastructure and resources are likely to be less in times of disaster. These include adequate emergency response teams, clean water supplies, equipment for treating and transporting the injured, and plans for recovery and psychological support and rehabilitation. We have to adders the importance of mental health nurses to be involved in short- and long-term mental health counseling and support immediacy post traumatic for the injured persons, the emotionally upset, the bereaved, and Post traumatic stress disorder is likely to be an ongoing health issue for surviving adults and children. Pediatric nurses and school nurses have a special role to play in caring for children in disasters Duong, K. (2009).

It is also critical to remember that nurses and their families may themselves be the victims of disasters. High anxiety states in health professionals may occur when disasters wallop, such as that we have seen occur in the 2002-3 SARS pandemic of the disaster when many nurses were infected or died from the virus, or had colleagues or families affected, emergency care response to disasters have only a limited impact on the health and recovery of affected communities, except, of course, at the level of those fortunate enough to receive life-saving treatment in a timely way. He contended that the real work to prepare for and recover from disasters needs to be undertaken by community members and health practitioners working in communities. Hence, primary health care and health promotion activities to prepare communities to mobilize resources for potential disasters are critical (Turale, s.2010).

Hospital Emergency Codes as Australian Codes:

- Code RED -- Fire
- Code GRAY -- Tornado or severe weather
- Code BLACK -- Bomb or bomb threat
- Code ORANGE -- Hazardous chemical, biological, or radioactive incident
- Code BLUE -- Medical Emergency
- Code WHITE -- Snow or weather emergency
- Code YELLOW -- Disaster, either internal or external
- Code GREEN -- Building evacuation
- Code BROWN -- Missing adult patient
- Code ADAM -- Missing or abducted child
- Code COPPER -- Communication Involving Utility Failure
- Code VIOLET- Violent Situation.

(http://www.sasvrc.qld.gov.au/SASVRC/Assets/Documents/Internal%20Emergency%20Response%20Plan.pdf

Leadership and disasters

Defibxitions of leadership fluctuated, but there are certain apparatus that are central to the perception of leadership. Leadership, first and chief, involves a process of interactions that occurs between the leader and his or her group. The second component of leadership involves influence , which is concerned with how a leader affects his or her followers. The third component is that it occurs in groups . Groups are made up of individuals that have a common purpose. The fourth and final component of leadership is concerned with accomplishing goals. It is the leader that is responsible for directing a group of individuals toward accomplishing some task or outcome. All of these component are directly interrelated with the deliberating the emergency and disaster plan in all phases of disaster management (Cotton, 2009; Stogdill, 1974). various ways in which nurses provide leadership during disasters. Nurses are leaders as the largest health care workforce, nurses provide leadership in clinical care for disaster victims at the scene of an emergency and their places of employment in the community, the hospital, or other locations. Nurses provide leadership at multiple levels including state, regional, national, and international.

Theories of Leadership

The earliest research about leadership was based on the study of men who were considered great leaders and who arose from aristocracy. This great man theory argues that a few people are born with the necessary characteristics to be great, and they can be effective leaders in any situation. The trait theory of leadership became popular in the mid-1940s and was based primarily on the "great man" theory, but it differed by taking the position that leadership qualities can be identified and taught (Marquis & Huston, 2005). Situational theory rose to prominence during the 1950s, expanding on "trait theory" with the caveat that leader traits vary and are determined by the particular situation (Sullivan, 1995). What was lacking in these early theories of leadership was the ability to predict which leadership behaviors would be most effective in specific situations. In situational leadership theory, the leader looks at the different variables surrounding the situation to make the best choice of leadership style. The leader will alter the style of leadership based on an analysis of the follower's readiness, meaning the level of motivation and competence an individual has for an assigned task. The leader assesses the follower's capacity to complete the assigned task and provides the appropriate leadership behavior that best meets the needs of the follower in the given situation.

Transformational leadership theory recognizes that for leaders to be effective, the organizational culture needs to change. With this leadership style, both the leader and the followers have the same purpose, and they raise one another to higher levels of performance. The transformational leader mobilizes others and grows and develops with the followers. In practice, establishing and maintaining both organizational and personal trust with others represents the fundamental strategy of the transformational leader. Integrative leadership theory concludes that the leader, the follower, and the situation all influence leadership effectiveness.

Leaders need to be aware of their own behavior and influence on others, they need to recognize the individual differences of their followers (characteristics and motivations), they need to understand the structures available to perform specific tasks, and they need to analyze the situational variables that impact the ability of followers to complete tasks, including environmental factors. With integrative leadership, the leader considers all of these factors using a "holistic" approach to oneself and others, and adjusts his or her leadership style through adaptive behavior. Similar to the evolution seen in the general theories of leadership, there has also been an evolution in the literature specific to nursing leadership. No longer is a "born" nurse leader viewed as effective in all situations (Sullivan, 1995).

There is growing consensus that different styles of nursing leadership are needed depending on the situation. As an example, in a crisis situation in which the followers have little or no knowledge or experience (e.g., patient in cardiac arrest), an autocratic style of nursing leadership may be most appropriate, where the leader takes total control, issues directives, and excludes group decision making (Sullivan, 1995). Alternatively, in non crisis situations, less autocratic styles may be appropriate. The thoughtful study and implementation of transformational leadership is recommended for implementation in health care organizations because it is congruent with nursing's values and organizational requirements. Accountability for practice is a hallmark of these organizations. Decision making and communication are shared equally in institutions that support transformational leadership (Sullivan, 1995).we need to take from every theory principals   what will support our leaders to be effective in the response phase of the disaster so we have to understand each theory exactly to apply it in our situation.

Disaster Leadership

Large-scale disasters starting in the late 1970s gave rise to much of the available literature on disaster leadership. These disasters included

• Tylenol's cyanide poisonings,

• National Aeronautics and Space Administration's Challenger and Columbia tragedies,

• Metropolitan Edison's Three Mile Island nuclear disaster,

• Exxon's Valdez oil spill,

• 2001 terrorist attacks, and

• Hurricane Katrina.

- 25th January revolution event and 30-6 chaos and terrorism that occur at emergency hospitals.

These events established crisis leadership as a new and rapidly developing field. A recent review found that 80% of existing literature has been published since 1985, establishing crisis leadership as a relatively new area of inquiry. Because of its recent development, the study of crisis leadership lacks an overall conceptual paradigm, is highly fragmented, covers a wide range of complex variables, and has been studied through the lens of numerous disciplines including economics, history, psychology and philosophy. Few data-based studies exist, and more analytical works result from attempts by researchers to synthesize anecdotal information. These analytic works are practitioner oriented rather than theoretical and are derived primarily from consultants hired to resolve crises for large corporations and governmental entities.

The definition of crisis leadership identifies three essential components: communication, clarity of vision and values, and caring relationships. Leaders who develop, pay attention to, and practice these qualities are better able to handle the important human dimension of a crisis. Crisis management focuses on planning; controlling; leading; organizing; and motivating prior to, during, or after a crisis. Crisis management is different than crisis leadership. With crisis leadership, the leader provides vision and influence in a non coercive manner to provide strategic decision making and guidance across the phases of the crisis. Crisis leadership includes crisis management but extends beyond to cultivate the followers' desire to achieve a vision and mission in a time of crisis (Porche, 2009 & Weiss, 2002).

When a disaster strikes, crisis leaders need to remind people of their strengths despite the fear and anxiety provoked by the event. By quickly offering information that is needed to make decisions, the crisis leader will empower people to help themselves and their loved ones survive. In a crisis, the crisis leader will be judged by the content of official messages, the speed of the communication, and the perception of their credibility. Leaders strongly influence the ability of individuals, organizations, communities, and nations to cope with and recover from crises. Reynolds, B. J. (2009).

Crisis leaders approach the public as equals in the disaster situation by empowering their health-risk decision making. Effective crisis leadership boils down to responding to the human needs, emotions, and behaviors caused by the crisis. Effective crisis leaders respond to the emotional needs perceived by those experiencing the crisis. People are more apt to follow a leader who is reassuring and who can meet their primary needs. Crisis leaders use integrative leadership and adapt the style of leadership to the given situation to ensure success. Comprehensive emergency management requires leaders to consider not only the immediate crisis response but also leadership for hazard mitigation and disaster preparedness and recovery. These components are intertwined and require a flexible leadership approach that is different than the approach needed during the actual emergency response. What this means is that a crisis leader needs to be cognizant of the possibility that other leadership approaches may be more effective during the different phases of a crisis. Sheetz, A. H. (2010).

Nursing leadership at multiple levels

There are various ways in which nurses provide leadership during disasters. From a preparedness perspective, nurses are leaders within their own homes for personal and family preparedness, assuring that their families have a disaster supply kit and a plan for responding to the threats that are likely to occur in their community. As the largest health care workforce, nurses provide leadership in caring for disaster victims whether that is caring for disaster victims at the scene of a disaster, providing care to the broader community, or caring for victims at their place of employment such as a health care facility. Nurses are increasingly providing leadership at the state, regional, national, and international levels where they lead planning and policy efforts to enhance the preparedness of the nation to respond to disasters. The history of nursing leadership in crises provides the context for understanding the ways in which nurses provide leadership at multiple levels. (Ann R. Knebel, Lauren Toomey, and Mark Libby, 2012).

The role of nurse leaders and administrators

Nurse leaders and administrators in hospital and critical and emergency settings, academic, and professional organizations have the great responsibility of leading and managing during times of crisis and disaster. Their role includes keeping open lines of communication, trying to maintain good patient care, providing education, influencing policy and financial decisions. During disasters clinical leaders have the difficult job of providing security and safe environments for staff, patients and families12. Nurse leaders in academic settings are often an untapped resource that can be utilized during disasters since they have well educated and trained staff and many students of nursing and midwifery at their disposal. Before disasters happen nurse leaders from all of these settings need to collaborate together and communicate with other stakeholders to make careful plans for utilizing resources during disasters. Clearly they need to ensure that they and their staff are educated and in a state of readiness, and can recognize potential resources that can be utilized when disaster wallops (Turale,2010)

Ethical challenges and evidence based and disaster management

A variety of ethical challenges are presented at the time of public health emergencies due to the fact that the hazard are often high since many people may be affected at once; there is diminutive time to deliberate decisions and solve any problems at this time of disaster situation and problem solve; and the urgent situation may have affected crucial resources such as roads, electrical power and so forth. the patterns of decision making they engaged in and possible improvements that could be made to improve their skills in these areas. The potential improvements can be applied to the use that nurses could make to them. Specifically, several studies concluded that potential improvements could be made in the areas of identifying a wider range of ethical issues to consider; by discussion and training, developing a deeper understanding of the ethical issues;

identifying and using resources to aid in identifying the issues and making decisions about them; assigning roles to designated persons and providing training for these people; reducing the vulnerability of the ethics environment when leadership turnover occurred and evaluating action taken in public health emergencies after they are over (Turale, 2010).

The potential improvements that these authors identified for their epidemiology section responders could easily be applied to the work of public health nurses. For example, nurses could have training including role playing, case studies and scenario development in order to identify the ethical dilemmas and work through possible solutions prior to a disaster. Nurses could also identify issues and possible responses, assign roles, design a care path that is not vulnerable to leadership changes and evaluate their actions in the face of a real or mock disaster so they would be better prepared to deal with the actual ethical challenges as they might arise from all evidence based practice in critical setting so we assert the importance of availability of evidence based practice in our hospitals although ethical consideration not changed but we have to develop and maintain it in our clinical setting , this can be supported by evidence based practice. (U.S. Department of Health and Human Services , 2000).

Reputation Repair and Behavioral Intentions

The second step is to review the intensifying factors of crisis history and prior reputation. If an organization has a history of similar crises or has a negative prior reputation, the reputational threat is intensified. A series of experimental studies have documented the intensifying value of crisis history and prior reputation. The same crisis was found to be perceived as having much strong crisis responsibility (a great reputational threat) when the organization had either a previous crisis or the organization was known not to treat stakeholders well/negative prior reputation. In general, a reputation is how stakeholder perceive an organization. A reputation is widely recognized as a valuable, intangible asset for an organization and is worth protecting. But the threat posed by a crisis extends to behavioral intentions as well. Increased attributions of organizational responsibility for a crisis result in a greater likelihood of negative word-of-mouth about the organization and reduced purchase intention from the organization. Early research suggests that lessons designed to protect the organization's reputation will help to reduce the likelihood of negative word-of-mouth and the negative effect on purchase intentions as well Hirshleifer (Coombs,J.2008).

Phases of disaster preparedness

Pre-Crisis Phase:

Prevention involves seeking to reduce known risks that could lead to a crisis. This is part of an organization's risk management program. Preparation involves creating the crisis management plan, selecting and training the crisis management team, and conducting exercises to test the crisis management plan and crisis management team. organizations are better able to handle crises when they (1) have a crisis management plan that is updated at least annually, (2) have a designated disaster management team, (3) conduct exercises to test the plans and teams at least annually, and (4) pre-draft some disaster messages. Table 1 lists the Crisis Preparation Best Practices. The planning and preparation allow disaster teams to react faster and to make more effective decisions (U.S. Department of Health and Human Services, 2009):

Disaster Preparation Best Practices

1. Have a disaster management plan and update and revised it at least annually.

2. Have a designate disaster management team that is properly trained.

3. Conduct exercise and drill at least annually to test the disaster management plan and team.

4. Pre-draft select disaster management communication messages

including content for dark web sites and templates for disaster statements. Have the legal department review and pre-approve these messages.

hospital preparedness assessment Checklist

Any item for which the answer is "no" will require action

- Indicate the action required.
- The steps taken to correct the condition and initial.
- Date the inspection checklist.

The item	YES	NO
• **1. Outdoor hazards**: • Railings, benches lights, poles all anchored well? • Overhanging trees branches trimmed? ○ **2. Building:** • No cracks/seepage visible in walls? • Basement dry; No likelihood of water? Compliance with codes (fire, etc)? ○ **3. Roof**: • Sloped or pitched (not flat)? • Covering sound? No cracks, no leaks? • Flashing, caulking intact? • Equipment on roof prohibited or properly anchored?		

○ **4. Windows and skylights**
• Caulking/sealants sound?
• Trees/limbs trimmed away?
•
 	
• **5. Fire Safety**
• Fire-resistant structure?
• Concrete floor/no air passages between floors?
  Concealed spaces (e.g. false ceilings) identified?
  Fire detection in concealed spaces?
  Stairways, pipe shafts enclosed?
  Electrical wiring in good condition?
  Appliance cords in good condition?
  Appliances unplugged nightly?
• Do Staff have keys to mechanical/janitorial rooms?
• Regular Fire Marshall visits?
 	
• Fire Marshall used productively? (Floor plans given to fire department? Priority areas noted? Appropriate follow-up on observed code violations?)
  Detection systems appropriate?
  Wired to 24 hour monitoring station?
  Tested regularly?
  Extinguishers present?
  Appropriate extinguishers? Inspected appropriately on schedule?
  Automatic suppression system? (present and operating?)
• Staff trained in:		

- Sounding alarms?
- Interpreting panels? (if present)
  Notifying Fire Department and others
  as called for?
- Locating in-house emergency
  equipment?
- Using fire extinguishers?
- Turning off power, gas, HVAC, etc.
- Closing fire doors?
- Supervising evacuation?

**6. Heating, ventilation and air
conditioning systems (HVAC)**

- Automatic shut-off in case of fire?
  Furnace/boiler inspected each year?
  Air Conditioning
- No leaks?
  No mold present?
  Effective drainage from pans?

- Dehumidification capacity?

- Able to run on exhaust to reduce
  smoke?

- **7. Stack Area**
- Shelves well braced?
- No water sources above collections?
- Files shelved snugly?
- Shelving 4-6" off the floor?
- "Canopies" atop shelving units?
  No valuable materials in basement?

• Exits unobstructed?		
• Important items away from windows?		
○ **9. Protection from water damage**		
• Pipes/plumbing well supported?		
• No pipe/plumbing leaks? Sump-pumps and backup present? No leaks/seepage from walls? Valuable materials stored above ground level?		
• Valuable/fragile media stored in protective enclosures?		
• Does staff have keys to mechanical rooms and janitorial closets? Does staff know location of water main, have appropriate tools for shut off, if needed?		
• Operable flashlights available?		
○ **9. Security**		
• Book drops away from building or in fire-resistant enclosure?		
• Building exterior well lighted?		
• Locks/alarms on windows and doors?		
• Intrusion detectors/alarms present and monitored 24-hours?		
• Effective closing procedures to ensure building is vacant?		
○ **10. Housekeeping**		
• Cleaning supplies and other flammables stored safely?		
• Trash removed nightly?		
• Staff room cleaned daily and well?		

• Smoking prohibited? • Food and drink prohibited? Prohibition enforced? • Pest management strategies in place and effective? ○ **11. Insurance:** • Policy up to date? • Replacement costs specified as needed? • Staff aware of records required for claim? Those records maintained safely? • Duplicate shelf list, catalog, and inventory and/or back-up computer tapes for entire collection? • **13. Construction projects**: • Responsibility for fire safety precautions clearly specified in contract? Fire guards used in all cutting and welding operations? • Debris removed nightly • Fire-resistant partitions used? Extra fire extinguisher on hand?		

Disaster Management Plan

A Disaster management plan (DMP) is a reference tool, not a blueprint outlines or proposal. A DMP provides lists of key contact information, reminders of what typically should be done in case of a disaster occurrence, and forms to be used to document the crisis response. A DMP is not a step-by-step guide or clinically pathway to how to manage a crisis or disasters. Than it is have noted how a DMP saves time during a crisis by pre-assigning some tasks, pre-collecting some information, and serving as a reference source. Pre-assigning tasks presumes there is a designated disaster team. The team members should know what tasks and responsibilities they have to do during a disaster which are illustrated in the disaster management plane.(John labby, 2010).

Disaster Management Team:

Barton (2001) identifies the common members of the disaster team as public relations, legal, security, operations, finance, and human resources. However, the composition and nature of this team will vary and show discrepancy based on the nature and type of the disaster. For instance, information technology would be required if the disaster involved the computer system. Time is saved because the team has already decided on who will do the basic tasks required in managing a disaster. notes that plans and teams are of little value if they are never tested. Management does not know if or how well an untested disaster management plan with work or if the disaster team can perform to potential outlook expectations. All the most recent research emphasized that training and well preparation is needed so that team members can practice making decisions in a crisis situation. As noted earlier, a DMP serves only as a rough guide. Each crisis is unique demanding that disaster teams make decisions. (U.S. Department of Health and Human Services, 2010)

Spokesperson

A key component of disaster team training is spokesperson training. Organizational members must be prepared to talk to the news media during a disaster. Media training should be provided before a disaster hits. The disaster Media Training Best Practices AS related to the references showed as the following: Spokesperson Media Training Best Practices:

1. Avoid the phrase "no comment" because people think it means the organization is guilty and trying to hide something
2. Present information clearly by avoiding jargon or technical terms. Lack of clarity makes people think the organization is purposefully being confusing in order to hide something.
3. Appear pleasant on camera by avoiding nervous habits that people interpret as deception. A spokesperson needs to have strong eye contact, limited disfluencies and avoid distracting nervous gestures such as fidgeting or pacing. Coombs (2007a) reports on research that documents how people will be perceived as deceptive if they lack eye contact, have a lot of disfluencies, nor display obvious nervous gestures.

Management team members consist of:

- Response Director: medical hospital director.
- Caller for external help and recorder : nursing manager
- Caller and alarm activator: medical administrative leader.
- Pack-Out and Relocation Supervisors: assistance of hospital manager
- Logistics Coordinator: Secretary-general of the hospital.
- Administrative Services Coordinator: Budget Officer
- Technical facilitator: Head of engineering department
- Computer Systems Manager: Head of I.T.
- Plan developer: Head of administration department
- Triaging team supervisor: Head emergency department.

The role of each team member:

- **Response Director:**

  - Determines scope of salvage operation and sets timetable for recovery.
  - Assigns specific tasks to specific disaster team members.
  - Organizes and supervises recovery teams and volunteers.
  - Establishing priorities.
  - Responsible for overall management of recovery and salvage operation.
  - Determines when to begin salvage after consulting with Physical Plant, Building representative, fire and safety.

- Establishes command center.

- Assesses and records damage.

- Determines the level of preservation response needed by consulting the Collection representative and priority lists.

- Directs Logistics Coordinator.

- Requests volunteers, as needed.

- Arranges training of crew team captains.

- Receives team reports.

- Prepares final report.

• Caller for external help and recorder :

- Identifies outside vendors as required

- Maintains a log of the recovery operation procedures.

- Keeps detailed record of damaged materials sent off-site or treated in faculty

- Provide personnel to carry out recovery plan. This may include removal of damaged materials from the disaster site; arranging materials for air-drying; packing materials for shipment off-site.

• Caller and alarm activator:

- Assess the situation and the capability to overcome.

- Acts as media representative for the Libraries.

- Give the needed instruction.

- Assess the response.

- Ask the people to be calm.

- **Pack-Out and Relocation Supervisors:**
  - Preliminary plans the suitable areas for evacuations.
  - Determine the safer and secure place for evacuations.
  - Supervise the process of evacuation.
  - Prepare a report on relocation activities which will include a photographic record.
  - Perform regular safety inspections of library facilities.
  - Supervise the training of volunteers in making and packing boxes.
  - Keep count of boxes.
  - Work with Collection Representative and keep general records of sections moved to other sites, depending on the size of the disaster;
  - Prepare a written report of the pack out activities;
  - Monitor the progress and orderly restoration of the stack area, including clean up and resetting shelving.
  - Organize and supervise the orderly return of faculty materials to approved shelving;
  - Keep records of the number of boxes and sections returned to the stacks.

- **Logistics Coordinator:**
  - Coordinates transportation & relocation activities
  - Arranges for the boxing, packing, air-drying and/or freezing of damaged materials

- Assists Coordinator as needed
- Maintains disaster recovery supplies
- Delegates functions as appropriate;
- Makes sure any volunteers sign waiver forms;
- Makes any necessary arrangements to remove books from the disaster site;
- Arranges for transportation and moving equipment;
- Supervises loading and unloading;
- Oversees shipping of boxes to freezers or other sites;
- Supervises delivery and installation of needed equipment;
- Supervises crews which set up the established recovery work place;
- Arranges the return of books to their original location;
- Coordinates with the appropriate building services and faculty staff.

  - **Administrative Services Coordinator:**
- Determines supply, equipment, and personnel needs
- Determines costs for recovery of damaged items
- Authorizes emergency expenditures
- Assesses damage to materials.
- Authorizes payment and signs vouchers for supplies and services needed for on-campus or outside vendors.
- Contacts vendors and services at the request of the disaster response director.
- Works closely with the logistics coordinator to arrange transport and delivery of needed supplies and services;

- **Technical facilitator:**
  - Provides expert knowledge of floor plans.
  - Will act as building reference person for emergency personnel.
  - Organizes emergency communication operations. e.g. temporary radio, phone, and data communications systems.

- **Computer Systems Manager:**
  - Check and calibrate Systems contact information
  - Ensures that damage to computers & software is minimized
  - Ensures that priority computers and applications are restored quickly

- **Plans developer :**
  - Assess all plants of the faculty
  - Assess the available mitigations in the faculty
  - Determine the mitigations needs.
  - Determine how to overcome the mitigations defect.
  - Develop the plans of disaster management including the four phases.
  - Develop the organizational structure of disaster management team.

Triaging team supervisor:

- Select the proper team members
- Gain their knowledge and practices regarding triaging approach.
- Check readiness occasionally to meet any disaster at any time.

Crisis Response:

The disaster response is what management does and says after the crisis hits. Public relations plays a critical role in the crisis response by helping to develop the messages that are sent to various publics. A great deal of research has examined the disaster response. That research has been divided into two sections: (1) the initial disaster response and (2) reputation repair and behavioral intentions. (Veenema T. 2009)

Initial Response

Practitioner experience and academic research have combined to create a clear set of guidelines for how to respond once a crisis hits. The initial crisis response guidelines focus on three points: (1) be quick, (2) be accurate, and (3) be consistent.

Be *quick* seems rather simple, provide a response in the first hour after the crisis occurs. That puts a great deal of pressure on crisis managers to have a message ready in a short period of time. Again, we can appreciate the value of preparation and templates. The rationale behind being quick is the need for the organization to tell its side of the story. In reality, the organization's sides of the story are the key points management wants to convey about the crisis to its stakeholders. When a crisis occurs, people want to know what happened. Crisis experts often talk of an information vacuum being created by a crisis.

The news media will lead the charge to fill the information vacuum and be a key source of initial crisis information. (We will consider shortly the use of the Internet as well). If the organization having the crisis does not speak to the news media, other people will be happy to talk to the media. These people may have inaccurate information or may try to use the crisis as an opportunity to attack the organization. As a result, crisis managers must have a quick response. An early response may not have much "new" information but the organization positions itself as a source and begins to present its side of the story. a quick response is active and shows an organization is in control. It lets others control the story and suggests the organization has yet to gain control of the situation. Arpan and Rosko-Ewoldsen (2005) conducted a study that documented how a quick, early response allows an organization to generate greater credibility than a slow response. Crisis preparation will make it easier for crisis managers to respond quickly.(Vinter S , Lieberman D A , Levi J : 2010).

Obviously ***accuracy*** is important anytime an organization communicates with publics. People want accurate information about what happened and how that event might affect them. Because of the time pressure in a crisis, there is a risk of inaccurate information. If mistakes are made, they must be corrected. However, inaccuracies make an organization look inconsistent. Incorrect statements must be corrected making an organization appear to be incompetent. The philosophy of speaking with one voice in a crisis is a way to maintain accuracy.

Speaking with one voice does not mean only one person speaks for the organization for the duration of the crisis. As Barton (2001) notes, it is physically impossible to expect one person to speak for an organization if a crisis lasts for over a day. Watch news coverage of a crisis and you most likely will see multiple people speak. The news media want to ask questions of experts so they may need to talk to a person in operations or one from security. The public relations department plays more of a support role rather than being "the" crisis spokespersons. The crisis team needs to share information so that different people can still convey a consistent message. The spokespersons should be briefed on the same information and the key points the organization is trying to convey in the messages. The public relations department should be instrumental in preparing the spokespersons. Ideally, potential spokespersons are trained and practice media relations skills prior to any crisis. The focus during a crisis then should be on the key information to be delivered rather than how to handle the media. Once more preparation helps by making sure the various spokespersons have the proper media relations training and skills.(Vinter S. , Lieberman, A , Levi J. 2010)

Quickness and accuracy play an important role in public safety. When public safety is a concern, people need to know what they must do to protect themselves. Sturges (1994) refer to this information as instructing information. Instructing information must be quick and accurate to be useful. For instance, people must know as soon as possible not to eat contaminated foods or to shelter-in-place during a chemical release. A slow or inaccurate response can increase the risk of injuries and possibly deaths. Quick actions can also save money by preventing further damage and protecting reputations by showing that the organization is in control. However, speed is meaningless if the information is wrong. Inaccurate information can increase rather than decrease the threat to public safety.

The news media are drawn to crises and are a useful way to reach a wide array of publics quickly. So it is logical that crisis response research has devoted considerable attention to media relations. Media relations allows crisis managers to reach a wide range of stakeholders fast. Fast and wide ranging is perfect for public safety—get the message out quickly and to as many people as possible. Clearly there is waste as non-targets receive the message but speed and reach are more important at the initial stage of the crisis. However, the news media is not the only channel crisis managers can and should use to reach stakeholders.

Web sites, Intranet sites, and mass notification systems add to the news media coverage and help to provide a quick response. Crisis managers can supply greater amounts of their own information on a web site. Not all targets will use the web site but enough do to justify the inclusion of web-base communication in a crisis response. extensive analysis of crisis web sites over a multiyear period found a slow progression in organizations utilizing web sites and the interactive nature of the web during a crisis. Mass notification systems deliver short messages to specific individuals through a mix of phone, text messaging, voice messages, and e-mail. The systems also allow people to send responses. In organizations with effective Intranet systems, the Intranet is a useful vehicle for reaching employees as well. If an organization integrates its Intranet with suppliers and customers, these stakeholders can be reached as well. As the crisis management effort progresses, the channels can be more selective. (Vinter S. , Lieberman A , Levi J. 2010)

More recently, crisis experts have recommended a third component to an initial crisis response, crisis managers should express concern/sympathy for any victims of the crisis. Victims are the people that are hurt or inconvenienced in some way by the crisis. Victims might have lost money, become ill, had to evacuate, or suffered property damage. Kellerman (2006) details when it is appropriate to express regret. Expressions of concern help to lessen reputational damage and to reduce financial losses. Experimental studies by Coombs and Holladay (1996) and by Dean (2004) found that organizations did experience less reputational damage when an expression of concern is offered verses a response lacking an expression of concern. Cohen (1999) examined legal cases and found early expressions of concern help to reduce the number and amount of claims made against an organization for the crisis. However, Tyler (1997) reminds us that there are limits to expressions of concern. Lawyers may try to use expressions of concern as admissions of guilt. A number of states have laws that protect expressions of concern from being used against an organization. Another concern is that as more crisis managers express concern, the expressions of concern may lose their effect of people. Hearit (2007) cautions that expressions of concern will seem too routine. Still, a failure to provide a routine response could hurt an organization. Hence, expressions of concern may be expected and provide little benefit when used but can inflict damage when not used.

Argenti (2002) interviewed a number of managers that survived the 9/11 attacks. His strongest lesson was that crisis managers should never forget employees are important publics during a crisis. The Business Roundtable (2002) and Corporate

Leadership Council (2003) remind us that employees need to know what happened, what they should do, and how the crisis will affect them. The earlier discussions of mass notification systems and the Intranet are examples of how to reach employees with information. West Pharmaceuticals had a production facility in Kinston, North Carolina leveled by an explosion in January 2003. Coombs (2004b) examined how West Pharmaceuticals used a mix of channels to keep employees apprised of how the plant explosion would affect them in terms of when they would work, where they would work, and their benefits. Moreover, Coombs (2007a) identifies research that suggest well informed employees provide an additional channel of communication for reaching other stakeholders.

When the crisis results in serious injuries or deaths, crisis management must include stress and trauma counseling for employees and other victims. One illustration is the trauma teams dispatched by airlines following a plane crash. The trauma teams address the needs of employees as well as victims' families. Both the Business Roundtable (2002) and Coombs (2007a) note that crisis managers must consider how the crisis stress might affect the employees, victims, and their families. Organizations must provide the necessary resources to help these groups cope.

We can take a specific set of both form and content lessons from the writing on the initial crisis responseForm refers to the basic structure of the response. The initial crisis response should be delivered in the first hour after a crisis and be vetted for accuracy. Content refers to what is covered in the initial crisis response. The initial message must provide any information needed to aid public safety, provide basic information about what has happened, and offer concern if there are victims. In addition, crisis managers must work to have a consistent message between spokespersons.

Initial disaster Response Best Practices:

1. Be quick and try to have initial response within the first hour.

2. Be accurate by carefully checking all facts.

3. Be consistent by keeping spokespeople informed of crisis events and key message points.

4. Make public safety the number one priority.

5. Use all of the available communication channels including the Internet, Intranet, and mass notification systems.

6. Provide some expression of concern/sympathy for victims

7. Remember to include employees in the initial response.

8. Be ready to provide stress and trauma counseling to victims of the crisis and their families, including employees.

Response to different type of disasters:

1-fire

A primary Fire Prevention involves a two fold approach to hospital fire safety by combining enforcement and regulation of the hospital fire code and hospital based education program, on the enforcement side, the Fire Office works closely with the hospital building code inspectors to ensure all new commercial and residential construction within the hospital meets the established fire code, this includes testing and approval of fire protection systems, plans review of new safety systems, fire alarms, fire extinguisher and reviewing architectural plans for "Life Safety Code" compliance. In addition, it oversees all efforts to bring all older commercial buildings up to present day code, the difference between how the city building code and the hospital fire codes function can be thought of like this; While the building code requires all new construction to meet current regulations requiring on site fire suppression systems, the fire code acts more like a maintenance regulation, this is done through annual inspection and maintenance.

Thousands of people annually suffer injuries caused by fires, the majority of fire related deaths 70 percent are caused by smoke inhalation of the toxic gases produced by fires, one third of residential fires caused by flammable products, such as gases, in addition, a lack of working smoke alarms can significantly increase the chance of dying in a residential fire. However, by taking appropriate steps to make your hospital safe, you can protect organization and the staff from fires by Keep flammable products saved and locked and out of the reach of any person, Install and maintain smoke and fire alarms, Keep and maintain your fire extinguishers, electrical Maintenance, regularly have electrical connections inspected, and turn off and unplug if not needed, a fireplace screen and Have your chimney cleaned and inspected yearly, develop several fire escape or evacuation plans from each room in the hospital and practice them regularly as a drill with your staff, make sure that is any items such as clothing or blankets do not cover lamps that are turned on, teach fire and burn safety behavior to your Staff Lastly, evaluation of the disaster plan's effectiveness should take place. All entities involved in the design of the community's disaster plan should come together in evaluating how well the plan met the needs of citizens. Evaluation data is important for improving disaster planning and preparedness (Exploratory Drilling in Amguri, 2004)

If you discover a fire inside a building:

Immediately implement **R.A.C.E.**:

- **Rescue** -- Rescue anyone in danger from the fire if it does not jeopardize your own life.

- **Alarm** -- Sound the alarm by calling police at **122** or by dialing the emergency number of the campus. Also activate a pull station to set off the building fire alarm.
- **Confine** -- Try to confine the fire by closing all doors and windows to trap the fire and slow its progress.
- **Extinguish** or **Evacuate** -- Extinguish the fire if possible and if you know how to use a fire extinguisher. Evacuate the area if the fire is too big to put out.

If you discover a fire outside a building:

- If you are on campus, contact UT police at **x-2600**. If you are off campus, call 9-1-1.
- Do NOT activate the building fire alarm system

If the fire alarm starts sounding:

- Feel the door or doorknob to the hallway with the back of your hand. If it feels hot, do not open it – the fire may be on the other side of the door.
- If the door is not hot, open it slowly. If the hallway is clear of smoke, walk to the nearest fire exit and exit the building. DO NOT USE ELEVATORS.
- Close doors behind you.
- Notify arriving fire or police personnel if you suspect someone is trapped inside the building, and where they may be located.

- Gather outside at a designated assembly area, and do not attempt to re-enter the building until instructed to do so by the Police or the Fire Department.

If you are trapped in a room, or otherwise unable to leave:

- Wet and place cloth material around and under the door to prevent smoke from entering the room.
- Close as many doors as possible between you and the fire.
- Be prepared to signal someone outside, but DO NOT BREAK GLASS until absolutely necessary (outside smoke may be drawn into the room).

If you are caught in smoke:

- Drop to hands and knees and crawl toward exit.
- Stay low, as smoke will rise to ceiling level.
- Breathe shallowly through nose and use a filter such as a shirt or towel.

If you are forced to advance through flames (which should be a last resort):

- Hold your breath.
- Move quickly.
- Cover your head and hair with a blanket or large coat.
- Keep your head down and your eyes closed as much as possible.

Using a fire extinguisher:

Building occupants are not required to fight fires. Individuals who have been trained in the proper use of a fire extinguisher and are confident in their ability to cope with the hazards of a fire may use a portable fire extinguisher to fight small fires (no larger than a waste paper basket).

Fire fighting efforts must be terminated within 15 seconds, or when it becomes obvious that there is risk of harm from smoke, heat or flames, which ever comes FIRST.

General fire fighter precautions:

The P.A.S.S. method:

Pull the safety pin from handle.
Aim the extinguisher at the base of fire.
Squeeze the trigger handle.
Sweep from side to side to side at base.

- Try to calm colleagues and patients.
- Disconnect the electricity, gas and air conditioning on incidence of fire.
- Divide personnel for work according to the available numbers.
- Collect files, documents away from the source of fire.

- Participate in transfer of patients to neighboring sections or places of evacuation reverse wind and fire.
- Can use the fire extinguisher and water hoses to (extinguish) douse the fire.
- Close all the doors and Windows that is located by fire place.
- Don't leave connecting doors open for not increasing the fire.
- Don't open door was the hot in grim.
- Put a banner on the door of a fire and no entry.
- determine safe evacuation locations for people to exit
- Transfer cases depending on the degree of risk for simple complex.
- Use the stairs not lifts to transfer the patients.
- Close the areas confirmed the evacuation and buster with the mark.
- Help the team to extinguisher fire.
- Provide first aid to the injured on the prioritized.
- Listing all people (the sick, visit nursing, doctors and workers) within the venue.
- Know from whom I take instructions at the time of fire.
- I know to whom I give instructions at the time of fire.

Hospital Police has the primary responsibility for managing fire emergencies with the Toledo Fire Department. Unauthorized re-entry into a building during a fire emergency is not permitted.

Violators of this policy are subject to University and state fire code sanctions.

Page top

- Prospective patients and staff
- Admission
- Academics
- hospital life
- Current Staff
- organization
- Research
- Athletics
- Community

EARTHQUAKES

The movement of the ground in an earthquake is rarely the direct cause of injuries; most are caused by falling objects or collapsing buildings. Many earthquakes are followed (several hours or even days later) by further tremors, usually of progressively decreasing intensity. To reduce the destructive effects of earthquakes a number of precautions are essential for people living in risk areas:

- Build in accordance urban planning regulations for risk areas.
- Ensure that all electrical and gas appliances in houses, together with all pipes connected to them, are firmly fixed.

- Avoid storing heavy objects and materials in high positions.
- Hold family evacuation drills and ensure that the whole family knows what to do in case of an earthquake.
- Prepare an emergency kit.

During an earthquake:
- Keep calm, do no panic.
- People who are indoors should stay there but move to the central part of the building.
- Keep away from the stairs, which might collapse suddenly.
- People who are outside should stay there, keeping away from buildings to avoid collapsing walls and away from electric cables.
- Anyone in a vehicle should park it, keeping away from bridges and buildings.

After an earthquake:
- Obey the authorities instructions.
- Do not go back into damaged buildings since tremors may start again at any moment.
- Give first aid to the injured and alien the emergency services in case of fire, burst pipes, etc.
- Do not go simply to look at the stricken areas: this will hamper rescue work.
- Keep emergency packages and a radio near at hand.
- Make sure that water and foods are safe to human use drink.

General care in case of earthquake:

- Disconnect the power immediately and any nearby gas and oxygen source.
- Disconnect the air conditioner of the place immediately.
- try restraint and calm of good behavior
- Contact immediately emergency number.
- Call the emergency team dealing with earthquakes.
- Avoid using any inflammable materials.
- Keep the doors open.
- Don't stand in balconies or roads section.
- Collect the patients and Panel in the middle of the section.
- Evacuate all people from port special emergency evacuation.
- Evacuate people from the place by priority.
- Help in organizing individuals and not accumulate them when exit from port emergency.
- Sit under a bed or table in case of inability to exit.
- try to stay away from things probably falling from above
- Remove any flammable materials such as smoking or oxygen cylinders.
- Do not enter the venue after exit until change ordered by the Director of the hospital.
- Know from whom I take instructions at the time of an earthquake.
- Know to whom I give instructions at the time of an earthquake.

Terrorism (civil disturbance):

Terrorism, The word *terror* derives from the Latin *terrere* meaning to *frighten*, kill one, terrorism or (civil disturbance) in the hospital is a threat or criminal act against hospital personnel, or property using the threat of violence and alter the behaviors of others, committed for political, ideological or religious purposes that creating mass anxiety, fear, and panic, fostering a sense of helplessness and hopelessness, demonstrating the incompetence of the authorities, destroying a sense of security and safety, with a desire to Manipulates the actions/reactions of others, Causes primary and/or secondary victimization and May result or does result in death, bodily injury or significant property damage, The intensity of the ripple effect depends on one's proximity to the event. Proximity can be measured geographically, emotionally, socially, politically, culturally, philosophically and financially, the indirect effects of terrorism are not always obvious.(Srinivas,(2005)

General Building/Office security:

- Don't prop open building/residence hall entrance doors/windows. Rectify these situations when you observe them.
- Don't leave keys unattended or give them to unauthorized persons.
- Immediately report lost keys or ID cards to the appropriate office.
- Secure all sensitive material/information when not able to attend to it.

- Secure sensitive deliveries in a timely manner.
- Secure all areas when not attended. Make sure doors to loading docks and emergency exits are not propped open.
- Be aware of unfamiliar persons in, or visitors to, office, etc.
- Protect access codes and combinations and change codes regularly. Report compromised codes immediately.
- Secure your laboratories, studios, darkrooms, and other areas people should not be in when unattended.
- Be sure to keep building doors and windows locked when the building is closed and not open for business.
- Do NOT allow tailgaters enter facilities behind you.
- Try restraint and calm to help the proper mindset.
- Feel responsible towards patients and not leaving or neglecting them.
- Call immediately the number of army or police.
- Contact nearby sections for assistance.
- Division of labor on the number of personnel available.
- Rushed the patient to evacuating area reverse event.
- Help the security to maintain the place and lock it.
- Disperse in place and not gather in one place.
- Close the doors from inside partition.
- Save important documents and files of patients quickly in a secure place.
- Share in transfer of precious devices and save it.
- Avoid the place by riot and inconvenience.
- Lie down on the ground in case of indiscriminate.

Crush Accident:

Any emergency hospital is Exposed for rapid increase flow rate of admission of large numbers patients such as crush accident and revolution patients, because of the injuries that sustained by a large numbers of people, newly explosions produce unique management challenges for health providers begin with an immediate surge of patients into surrounding health care facilities, the potential for the large numbers of patients arriving within a few hours may stress and limit the ability of emergency medical services systems, hospitals, and other health care facilities to care for critically injured victims.(Ashkenazi, I. ,et al 2010)

When the quantity and severity of injuries overwhelm the operative capacity of health facilities, different approach to medical treatment must be adopted, triage consists of rapidly classifying the injured on the basis of the severity of their injuries and the likely hood of their survival with prompt medical intervention, it must be adopted to locally available skills, higher priority is granted to victims whose immediate or long-term prognosis can be dramatically affected by simple intensive care. Moribund patients who require a great deal of attention, with questionable benefit, have the lowest priority, triage is the only approach that can provide maximum benefit to the greatest number of injured in a major disaster situation (Park, 2005)

The most common classification uses the internationally accepted four colures code system. Red indicates high priority treatment or transfer, yellow signals medium priority, green indicates ambulatory patients and black for dead or moribund patients, Triage should be carried out at the site of disaster, in order to determine transportation priority, and admission to the different departments of the hospital, where the patient's needs and priority of medical care will be reassessed. Ideally, local health workers should be taught the principles of triage as part of disaster training. Persons with minor or moderate injuries should be treated / at their own homes, the seriously injured should be transported to hospitals with specialized treatment facilities, all patients should be identified with tags stating their name, age, place of origin, triage category, diagnosis, and initial treatment, the major challenges that hospitals will face in a Mass Casuality Event (MCE) include surge capacity and capability issues in emergency and trauma services, as well as medical, paramedical, administrative, logistical, and security challenges, difficult decisions will have to be made regarding the allocation of available resources, these decisions should reflect circumstances and local regulations. So we have to follow certain instructions when dealing with this type of disasters in reference to (Wilson, H. 2010). :

- Contact immediately the emergency number to call emergency team.
- Contact all emergency departments for assistance.

- Setup the general condition of the patient through the examination of the general health status of the patient.

- Detect the level of consciousness of the patient.

- Split the cases in the incidents depending on four colors, known as the (red/yellow/green/black).

- Identify the type and priority of nursing care provided for the patient.

- Classify the patients according to type of incident.

- Help in the transfer of the patient to the emergency room as classified.

- Provide immediate care for injuries that threaten the life of the patient, such as cardiac and respiratory arrest.

- Provide nursing care during the first ten minutes of the impending injury which threatens the life of the patient, such as hemorrhage and declining circulation.

- Provide nursing care during the first half hour of the possible injuries that threaten the life of the patient, such as simple bleeding and simple burns.

- Take the patient's vital signs (breathing, pulse, blood pressure and temperature).

- Document vital signs and care provided to patient for follow-up and recurrence prevention

- Examine the patient jaw injuries for head injuries.

- Maintain contact with other team members in the emergency rooms.

- Maintain the patient privacy.
- Implement required medical instructions to the patient.
- Prepare and assess patient awaiting medical round and investigation.
- Register and inventory all cases in the records.
- Know from whom I take instruction when entering the victims.
- Know for whom I give instruction When entering the victims.

EVACUATION -- An organized withdrawal from a building or area to reach safe haven.

Prepare:

Determine in advance the nearest exit from your work location, classroom or dorm room, and the route you will follow to reach that exit in an emergency. Establish an alternate route to be used in the event your route is blocked or unsafe.

During Evacuation:

- Evacuate quickly.
- Follow instructions from emergency personnel and follow the directions provided for safe routes of evacuation.
- Check doors for heat before opening. (Do not open door if hot).
- Close the door as you exit your room or office.
- Dress appropriately for the weather.

- Take only essentials with you (e.g., eyeglasses, medications, identification and cash/checkbook/credit cards) - do not pack belongings.
- Turn off unnecessary equipment, computers and appliances.
- Walk, do not run. Do not push or crowd.
- Keep noise to a minimum so you can hear emergency instructions.
- Use handrails in stairwells; stay to the right.
- Assist people with disabilities.
- Listen to a radio, if available, to monitor emergency status.

If you are relocating outside the building:

- Move quickly away from the building.
- Watch for falling glass and other debris.
- Try to stay with your fellow employees so all can be accounted for.
- If you have relocated away from the building, DO NOT return until notified that it is safe to do so.

SHELTER IN PLACE

During certain emergency situations, particularly chemical, biological or radioactive material releases and some weather emergencies, you may be advised to "shelter in place" rather than evacuate the building.

When directed to shelter in place:

- Stay inside the building (or go indoors as quickly as possible).

- Do not use elevators.

- Quickly locate supplies you may need such as food, water, radio, etc.

- If possible, go to a room or corridor where there are no windows and few doors.

- If there is time, shut and lock all windows and doors, (locking the door may provide a better seal on the door against chemicals).

- In the event of a chemical release, go to an above ground level of the building; some chemicals are heavier than air and may seep into basements even if windows are closed.

- Turn of the heat, fans, air conditioning or ventilation system, if you have local control of the systems.

- Drink bottled water or stored water, not water from the tap.

- If possible, check for additional information through the local radio and television stations, or on campus or government's website.

- Do not **call 122** unless you are reporting a life-threatening situation.

- If you smell gas or vapor, hold a wet cloth loosely over your nose and mouth and breath through it in as normal a fashion as possible.

When the "All Clear" is announced:

- Open windows and doors.
- Turn on heating, air conditioning or vent system.
- Go outside and wait until the building has been vented.

Post-Crisis Phase

In the post-crisis phase, the organization is returning to business as usual. The crisis is no longer the focal point of management's attention but still requires some attention. As noted earlier, reputation repair may be continued or initiated during this phase. There is important follow-up communication that is required. First, crisis managers often promise to provide additional information during the crisis phase. The crisis managers must deliver on those informational promises or risk losing the trust of publics wanting the information. Second, the organization needs to release updates on the recovery process, corrective actions, and/or investigations of the crisis. The amount of follow-up communication required depends on the amount of information promised during the crisis and the length of time it takes to complete the recovery process. If you promised a reporter a damage estimate, for example, be sure to deliver that estimate when it is ready. West Pharmaceuticals provided recovery updates for over a year because that is how long it took to build a new facility to replace the one destroyed in an explosion. Intranets are an excellent way to keep employees updated, if the employees have ways to access the site. Mass notification systems can be used as well to deliver update messages to employees and other publics via phones or cell phone (Veenema T. 2009).

Crisis managers agree that a crisis should be a learning experience. The crisis management effort needs to be evaluated to see what is working and what needs improvement. The same holds true for exercises. Every crisis management exercise be carefully dissected as a learning experience. The organization should seek ways to improve prevention, preparation, and/or the response. As most books on crisis management note, those lessons are then integrated into the pre-crisis and crisis response phases. That is how management learns and improves its crisis management process. Table 8 lists the Post-Crisis Phase Best Practices. (U.S. Department of Health and Human Services2010).

Post-Crisis Phase Best Practices

1. Deliver all information promised to stakeholders as soon as that information is known.

2. Keep stakeholders updated on the progression of recovery efforts including any corrective measures being taken and the progress of investigations.

3. Analyze the disaster management effort for lessons and integrate those lessons in to the organization's crisis management system. Disaster Action Team Handbook, (2008),

1- Mitigations in the hospital.

Mitigation in the hospital should be established to decrease the likelihood that an event or crisis will occur and to eliminate or reduce the loss of life and property damage related to an event or crisis. It's a continuous process (lifelong) including the following protection activities:

- Locks on doors and windows are secure and all keys accounted for.
- No pipes, faucets, water fountains, toilets, or air conditioning units are leaking.
- Electrical equipment is turned off or unplugged if not in regular use, and no frayed wiring is in evidence.
- There are no signs of structural damage.
- Combustible or burning materials are discarded in appropriate containers.
- Areas known to be problem locations should be checked often.
- Obtain funding for development and distribution of information and education plans for responding to all hazards.
- Develop distribution centers on campus where all-hazard
- Information and safety guidance can be provided to faculty staff, and students.
- Making educational materials for all hazards.
- Establish radio communication system.

- Evaluate and upgrade warning systems.
- Review and evaluate the community fire alarm system.
- Create lab safety training class for graduate students.
- Create a hardened vault in the lab for highly valuable lab equipment such as microscopes.
- Testing and approval of protection systems e.g. automatic closing doors.
- Alarm system. e.g. sprinklers, smoke detectors.
- Coding system. e.g. red for fire, blue for earthquake.
- Non-combustible materials. e.g. heat resistance doors lock.
- Proper ventilation windows.
- Evacuation stairs.
- Disaster Trunks There are trunks containing disaster supplies and it should be in well known and easy to reach place, each trunk contains the following emergency

- **Required supplies:**

  Batteries
  Bucket
  Disaster Plan
  Extension cord
  Flashlight
  Gloves – latex
  Gloves – rubber
  Jackets – tyvek
  Markers
  Newsprint paper
  Paper towels – folded
  Paper writing pads

Pencils
Pens
Plastic sheeting
Safety goggles
Scissors
Sponges
Tape – strapping

2- Preparedness activities:

Preparedness is the second phase in disaster management process, it's continuous activities includes the establishing the plan of response, recovery and training of faculty staff prior to a disaster to carried out the disaster management plan efficiently and smoothly.

- Establish organizational structure for disaster management team illustrates the line of authority.

- Set up emergency two different evacuation floor marshals and plans for the faculty with detailed information.

- Determine the emergency service providers, and vendors contact numbers.

- Develop faculty preparedness assessment checklist that assess the ability of the faculty to confront any type of disasters.

- Select response, recovery, and triaging team members often ready to confront any type of disaster.

- Develop the training program and always updates for previous mentioned teams.

- Provide disaster response instructions for all staff written and in seminars.

- Establish periodically drills to ensure the capabilities of building and humans to meet the different type of disasters.

- Check the signs that guide evacuations pathways.

- Develop alarm coding system.

3 – The response phase of the disaster:

The phase starting by disaster occurrence up to stoppage the threatening event e.g. (fire), it's the most threatening time peoples become panic and difficult to control. Good mitigation and preparedness lead to minimize the incidence, time, hazards of disaster. The disaster response clinical pathway summarizes, organize, and prioritize the response activity.

The basic principles of managing the response to a disaster are:

- Reporting call 122.
- Activate manual alarm if available.
- Advise your name and classification.
- Location of the event.
- Do not shout or panic, this may cause confusion.
- The response director assesses the situation and call vendor if needed.

- If fire and smoke by closing doors and, where practical, windows. This action will localize fire and smoke and reduce spread.
- Remove people from danger as quickly as possible.
- Prevent other people inadvertently coming into a danger area
- Minimize the damage to the physical structure of the faculty
- Maintain role and re-establish services.
- Administer first aid.
- DO NOT move the seriously injured unless in danger.
- Turn off utilities if applicable.
- Evacuate departments to safe areas. Record the people who are in or who have been evacuated from the building. If evacuated, record where to and time of evacuation.
- Don't use the lifter.
- In mass casuality the triaging team should be starting the Triaging until ambulances arrival.
- Immediate search and Rapid Triage.
- Immediate First Aid Treatment.
- Transport to the hospital.

4- Recovery phase:

The last phase starting from removal the risk e.g. fire up to regain the normal function of the institution.

- Assess the damage by the recover team.

Determine the quantity of materials to be salvaged based on the damage assessment by the Disaster Director and established collection salvage priorities; decide which materials need to be freeze dried and which can be air dried; decide if faculty Preservation can manage the disaster or if outside assistance is necessary.

- Disaster Response Director (chooses the site) and Logistics Coordinator (sets up the site).
- Organize Staff and Volunteers
- begin wrapping and packing damaged materials; all able staff, including student employees, may be asked to participate in salvage teams; the Disaster Response Director may also request help from volunteers; volunteers will sign a waiver of responsibility before beginning work; all must be given breaks and food in case of a major disaster.
- Arrange for Transportation, Supplies and Equipment.
- Logistics Coordinator and Administrative Services Coordinator
- Arrange for the delivery of materials and equipment as needed to the disaster site (including milk boxes, cardboard boxes,

garbage bags, wrapping paper, book trucks, dollies, flat trucks, labels and pens) from on-site or off-site sources; arrange for transportation of boxed damaged items from the disaster site to the recovery area; assemble any additional supplies or equipment, such as tables, chairs, lights, fans, as needed, especially for on-site salvage in a pre-determined, large, clean area with delivery access (Magnaye, Steffi , AnnF. , Gilbert. & Heather, 2011)

- **Pack-Out and Damaged Material**
- Pack-Out Supervisors meet with the Disaster Response Director and are briefed on the situation.
- Teams assemble and begin the packing procedure according to preservation-approved techniques.
- During packing, ranges and number of boxes are recorded.
- When trucks arrive, the Pack-Out Supervisors brief the moving crews and oversee the loading of pallets.
- Pack-Out Supervisors oversee the unloading at the recovery site.
- Supervisors report regularly on progress to the Disaster Response Director.
- Pack-out of Damaged Materials.
- Identify and secure before packing begins:
- Work space (air-drying location, freezer, and storage space).
- Transportation (arranged by the Logistics Coordinator).
- Packing area, with room to sort and pack materials.

- Loading area for receipt of supplies and shipping of wet books.
- Route for the removal of full boxes.
- Rest area for workers.
- All responsible will reporting the director the details of event (Cox & Briggs 2004).

DISASTER RESPONSE INSTRUCTIONS FOR ALL HOSPITAL MEMBERS:

Follow these instructions if you are first to discover an emergency situation.

1. REPORT THE EMERGENCY

Life-threatening emergency:

- Pull the fire alarm
- Leave the building (Public Safety Officers responding are responsible for evacuation)
- Notify hospital manager.

In reporting the emergency, remain calm, and provide the following information:

- Nature and location of the problem
- When the problem started
- Your name, location, and phone number
- Stay on the phone until you have given all the necessary information. Do not be the first to hang up.

2. ENSURE SAFETY

Do not enter the affected area until it has been determined that is safe to do so. If the **Fire Department** is called, they will be in charge of deciding when staff may re-enter the building. In a **Water** emergency, potential dangers include electrical shock, and exposure to sewage, chemicals, and mold.

To ensure safety in a **Water Emergency**:

- Turn off electricity in affected areas if necessary.
- Turn off water supply if it continues to run into affected areas.
- Ensure that no chemicals, sewage, or mold present health risks
- Cordon off unsafe areas

Contaminated Water

If you are responding to a **water emergency** but have not determined the source, remember that the water could be contaminated. If you are handling affected collections or working in the wet area, wear protective clothing. Rubber gloves, safety glasses, and protective jackets are in the Disaster Trunks located in Circulation, the Mailroom, and outside the Women's Room near Preservation on the Lower Level (Masellis, Ferrara, & Gunn, 1999).

3. HALT DAMAGE TO COLLECTIONS

In a **Water Emergency**, turn off the water supply if you can identify the source and can shut it off, AND Shield faculty materials from the source of water by:

- Covering faculty materials with plastic sheeting if water is coming from above
- Making a dam barrier to keep water away from faculty materials if flooding is from below.
- Moving undamaged materials to another location if they are in danger.

• References:

Ann R. Knebel, Lauren Toomey, and Mark Libby(2012), Nursing Leadership in Disaster Preparedness and Response, Springer Publishing Company

Ardalan, A., Mowafi, H., & Homa Yousefi, K. (2013). Impacts of natural hazards on primary health care facilities of Iran: A 10-year retrospective survey. PLoS currents, 5.

Attorney Robert Smith (2005) , Critical-thinking skills and a systematic approach can remedy many potential pitfalls. applies the phrase layers of safety in relation to telephone triage practice.

Baack, S. (2011) analysis of Texas nurses" preparedness and perceived Competence in managing disasters, The University of Texas at Tyler,pp 66-69.

Cox, E. & Briggs, S. (2004)progressive and critical care, by the American Association of Critical Care Nurses,Critical Care Nurse, Journal of high acuity, vol. 24 no. 3 PP: 16-2274

Crisis management workbook, (2010), Fairfax County Public Schools, The Office of Safety and Security, pp:234-238, available at: http://www.fcps.k12.va.us/DGS/safety security/planning/cmw.pdf

Disaster Action Team Handbook, (2008) American Red Cross, regional chapter, south plains, PP: 507-512. Available at: http://www.redcross.org/services/disaser/0,1082,0_601_,00.html

Disaster Action Team Handbook, (2008), American Red Cross Effective, south plain regional chapter An Annex to the Volunteer Handbook

Doyle, M. (2009) Terrorism preparedness: Perceptions of connectivity of emergency nurses of the Emergency Nurses Association, Theses, Dissertation, pp.:531–537.

Duong, K. (2009)"Disaster education and training of emergency nurses in Australia," AENJ, vol. 12, pp. 86-92.

Dybdahl,R. (2001), Children and mothers in war: an outcome study of psychosocial intervention program, Child Dev, PP;72 :80.

Eric,K. (1997), the Public Health Consequences of Disasters, NewYork: Oxford University Press, PP 213-214.

Exploratory Drilling in Amguri, (2004), Disaster Management Plan, Envirotech Consultants Pvt. Ltd.India.

Fahlgren,T. & Drenkard, K. (2002) Healthcare systems disaster preparedness, part 2nd edition . JONA, P. 32. pp:531–537.

Fairfax County Public Schools (2010) The Office of Safety and Security, Crisis management workbook, available at:http://www.who.int/ar/index.html.

Hirshleifer, CoombsJ. (2008). "Disaster and Recovery". In David R. Henderson (ed.). Concise Encyclopedia of Economics (2nd ed.). Indianapolis: Library of Economics and Liberty. ISBN 978-0865976658. OCLC 237794267.

John labby, (2010). Disaster Management, University of Taxes, Provided by Environmental Health and Safety.

Joseph Mc. ,Sandesh S. (2010). Emergency procedures – flowcharts construct an emergency procedure flowchart. www.scribd.com/doc/40701894/Emergency-Procedures-Flowcharts-1

Korstanje, M. (2011). "Swine Flu, beyond the principle of Reisilience". International Journal of Disaster Resilience in the Built Environment, Vol. 2 Iss: 1, pp. 59–73

Magnaye, B., Steffi,M., AnnF. , Gilbert, V. & Heather, M. (2011). The Role, Preparedness, and Management of Nurses during

Disasters. College of Nursing and Health Sciences. The University of Texas at Tyler PP 221-223

Masellis,M., Ferrara, M. &Gunn,S. (1999) Fire Disaster And Burn Disaster: Planning and Management, Palermo Mediterranean Club for Burns and Fire Disasters –the WHO Collaborating Centre Annals of Burns and Fire Disasters -vol. XII -n° 2 pp67-69.

McConell, C., Leeming, F., & Dwyer, W. (1996) „Evaluation of a fire safety-training program for preschool children, pp. 213-227

McMillan, J. (2004). Educational Research:FUndamental for the Consumer (4th ed).Boston: Pearson education, Inc, pp. 546-555.

Murat, C., Kevser, V., Harun, B., Cavit, I. & Ozlem, S. (2008). Introduction to Disaster Management, The Disaster Management Cycle. Available at: http://www.gdrc.org/uem/disasters/1-dm_cycle.htm.

National Clearinghouse for Educational Facilities at the National Institute of Building Sciences, (2009) www.ncef.org Prepared under a grant from the U.S. Department of Education, Office of Safe and Drug-Free Schools

Park, K. (2005) Preventive and Social Medicine, 18th ed, Jabalpur; Banarsidas Bhanot publishers; pp600-605

Reynolds, B. J. (2009). An exploration of crisis experience and training as they relate to transformational leadership and rhetorical sensitivity among U.S. public health offi cials . Retrieved from PsycINFO database. (UMI No. 3344533)

Sheetz, A. H. (2010). The H1N1 pandemic: Did school nurses assume a leadership role? Some thoughts on leadership. NASN School Nurse, 25 (3), 108–109.

Srinivas, H. (2005) Disasters: a quick FAQ. Accessed on 24/01/08 available at: http://www.gdrc.org/uem/disasters/1-what_is.html

Turale, s.(2010)Nurses: Are We Ready for a Disaster? Yamaguchi University, Ube, Yamaguchi, JapanJ Nurs Sci Vol.28 No.1 Jan - Mar 2010 Journal of Nursing Science

U.S. Department of Health and Human Services, (2000): Training manual for mental health and human ser-vices workers in major disasters, ed 2 , Washington, DC, (SAMHSA). Available at: http://mentalhealth.samhsa.gov/dtac/ccptoolkit/

U.S. Department of Health and Human Services(2010) : Healthy People: a roadmap to improve all Americans Health, Washington, DC , 2010 , US Government Printing Office . Available at http://www.healthypeople.gov. Accessed February 26, 2010

U.S. Department of Health and Human Services, (2009): National Health Security Strategy of the United States of America, Washington DC , 2009 , USDHHS , Available at http://www.hhs.gov/aspr/opsp/nhss/index.html. Accessed January 22, 2010

Uscher Pines, L. (2009)."Health effects of Relocation following disasters: a systematic review of literature". Disasters. Vol. 33 (1): 1–22.

Veenema T G: Ready RN, (2009): The healthcare technology readiness source for disaster and emergency preparedness staff, New York , Tener Consulting Group LLC, Elsevier Publishing, and MC Strategies Available at http://www.readyrn.com/. Accessed March 6, 2010.

Vinter S , Lieberman D A , Levi J : (2010)Public health preparedness in a reforming health care system Harv Law Policy

Rev4:339 – 360 ,. Available at http: // healthyamericans .org / assets / files / HLPR _ TFAH . pdf Accessed August 2, 2010 .

Wilson, H. (2010). "Divine Sovereignty and The Global Climate Change debate". Essays in Philosophy. Vol. 11 (1): 1–7.

Close

Disasters one of destructive barrier that prevent the human and communities to develop, Health Care Centers should be highly prepared and ready to meet any type of disaster effectively and efficiently since the first asylum for harmed persons in addition to the dependency of stakeholders or patient, the Disaster Management is Avery complicated process and need many and many supportive researches to alleviate the capability of institutions and persons to manage the difficult situations, I hope this product add to readers and the gain the field of interest.

Notes

www.ingramcontent.com/pod-product-compliance
Lightning Source LLC
Chambersburg PA
CBHW022120170526
45157CB00004B/1700